GASOHOL

A TECHNICAL MEMORANDUM

University Press of the Pacific
Honolulu, Hawaii

Gasohol:
A Technical Memorandum

by
Office of Technology Assessment

ISBN: 1-4102-2096-6

Reprinted from the 1979 edition

University Press of the Pacific
Honolulu, Hawaii
http://www.universitypressofthepacific.com

PREFACE

The Office of Technology Assessment is currently preparing an assessment of energy from biological processes. In the course of this study we have carried out an extensive analysis of alcohol fuels from agricultural products. This technical memorandum presents these findings in response to congressional interest in synthetic fuels. The purpose of the memorandum is to illuminate the technical and non-technical issues surrounding the development of gasohol. It discusses the resource base, production technologies, and economics of gasohol, and its use as a transportation fuel. The report also contains a discussion of the environmental problems and benefits of producing and using gasohol, and the social and institutional issues about using agricultural products for energy.

While the memorandum does not present an analysis of policy issues, it does provide estimates of how much gasohol can be used at what cost, and the long-term prospects for ethanol production. All are important to the current congressional debate over development of a gasohol policy.

The final report on energy from biological processes is scheduled for delivery to Congress in January 1980 and will contain an analysis of policy options about gasohol as well as other bioenergy technologies such as wood and methanol production.

JOHN H. GIBBONS
Director

Advisory Panel

Energy From Biological Processes

Dr. Thomas Ratchford, Chairman
Associate Executive Director
American Association for the
  Advancement of Science
Washington, D.C.

Dr. Henry Art
Center for Environmental Studies
Williams College
Williamstown, Massachusetts

Dr. Stanley Barber
Department of Agronomy
Purdue University
West Lafayette, Indiana

Dr. John Benemann
Sanitary Engineering Laboratory
University of California
Richmond Field Station
Richmond, California

Dr. Paul F. Bente, Jr.
Executive Director
The Bio-Energy Council
Washington, D.C.

Mr. Calvin Burwell
Oak Ridge National Laboratory
Oak Ridge, Tennessee

Dr. Robert Hirsch
EXXON Research and Engineering
  Company
Florham Park, New Jersey

Mr. Robert Hodam
California Energy Commission
Sacramento, California

Mr. Kip Howlett
Georgia Pacific
Washington, D.C.

Mr. Ralph Kienker
Monsanto Company
St. Louis, Missouri

Mr. Dean Kleckner
President, Iowa Farm Bureau
  Federation
West Des Moines, Iowa

Mr. Kevin Markey
Friends of the Earth
Denver, Colorado

Mr. Jacques Maroni
Energy Planning Manager
Ford Motor Company
Dearborn, Michigan

Dr. Michael Neushul
Marine Science Institute
University of California at
  Santa Barbara
Santa Barbara, California

Dr. William Scheller
Department of Chemical
  Engineering
University of Nebraska
Lincoln, Nebraska

Mr. Kenneth Smith
Office of Appropriate Technology
Sacramento, California

Dr. Wallace Tyner
Department of Agricultural
  Economics
Purdue University
West Lafayette, Indiana

NOTE: The Advisory Panel provides advice and comment throughout the assessment, but the members do not necessarily approve, disapprove, or endorse the report for which OTA assumes full responsibility.

ENERGY FROM BIOLOGICAL PROCESSES STAFF

Lionel S. Johns, Assistant Director

Energy, Materials, and Global Security Division

Richard E. Rowberg, Energy Group Manager

Thomas E. Bull, Project Director

A. Jenifer Robison, Assistant Project Director

Mark Gibson, Federal Programs

Steven Plotkin, Environmental Issues

Richard Thoreson, Economic Issues

Lisa Jacobson   Lillian Quigg   Yvonne White

SUPPLEMENTS TO STAFF

Stanley Clark

Robert Vernon

ACKNOWLEDGEMENTS

This document was prepared in conjunction with OTA's assessment of Energy from Biological Processes. In addition to reviews by the members of the Advisory Panel for Energy from Biological Processes, this technical memorandum has also been reviewed by the following people, but they do not necessarily approve, disapprove or endorse the report.

Mr. Weldon Barton
Office of Energy
U.S. Department of Agriculture
Washington, D.C.

Prof. Carroll Bottum
Department of Agricultural Economics
Purdue University
West Lafayette, IN

Prof. Otto Doering
Department of Agricultural Economics
Purdue University
West Lafayette, IN

Prof. Irving Goldstein
Department of Wood and Paper Science
North Carolina State University
Raleigh, NC

Mr. Sanford Harris
Chief, Biomass Branch
Division of Solar Technology
Department of Energy
Washington, D.C.

Ms. Marilyn Herman
Chairperson, Alcohol Fuels Policy Review
Office of Policy and Evaluation
Department of Energy
Washington, D.C.

Prof. Edward Hiler
Department of Agricultural Engineering
Texas A&M University
College Station, TX

Prof. Arthur Humphrey
Department of Chemical and
Biochemical Engineering
University of Pennsylvania
Philadelphia, PA

Prof. Ronald Lacewell
Department of Agricultural Economics
Texas A&M University
College Station, TX

Dr. Edward Lipinsky
Battelle Columbus Laboratories
Columbus, OH

Dr. Dwight Miller
Assistant Center Director
Northern Regional Research Center
USDA
Peoria, IL

Mr. Edward Noland
General Electric Company
Philadelphia, PA

Prof. Richard Pefley
Department of Mechanical Engineering
Santa Clara University
Santa Clara, CA

Prof. Roy Sachs
College of Agricultural and
Environmental Sciences
University of California
Davis, CA

Dr. Frank Sprow
EXXON Research and Engineering Co.
Florham Park, NJ

Dr. R. Thomas Van Arsdall
Office of Energy
U.S. Department of Agriculture
Washington, D.C.

TABLE OF CONTENTS

Issues and Findings

Gasohol is a mixture of one part ethanol (commonly known as "grain alcohol" or beverage alcohol) and nine parts unleaded gasoline. The ethanol can be produced from several types of plant material using technology that is currently available, and in most cases gasohol can be substituted for gasoline with only minor changes in mileage and performance. Another type of gasohol can be produced with one part methanol ("wood alcohol" or methyl alcohol) and nineteen parts gasoline. Although a consideration of this fuel is not included here, methanol fuel will be included in the later report on Energy from Biological Processes.

This memorandum addresses the major technical, economic, environmental and social factors related to gasohol production and use. It also contains a summary of the current federal gasohol programs and policies but does not include an analysis of policy options. The most important points developed in this report are summarized below as brief discussions of crucial questions.

o   WILL USING GASOHOL SAVE GASOLINE?

The amount of premium fuels (oil and natural gas) displaced by ethanol depends critically on the boiler fuel used at the ethanol distillery and the way the ethanol is used in gasohol. The distillery producing most of the fuel ethanol today uses natural gas as a boiler fuel and the ethanol is used to produce a high octane gasohol. Because local conditions enable this distillery to be particularly energy efficient, the use of gasohol made this way currently saves 1/3 gallon of gasoline and natural gas energy equivalent for every gallon of ethanol (i.e., for every ten gallons of gasohol). Less energy efficient distilleries fueled with premium fuels, however, could result in a net increase in premium fuel usage with gasohol production.

If distilleries are fueled with coal and solar energy (including biomass), and the ethanol is used to produce a "regular"-grade gasohol, however, the energy balance is far more attractive. Under these more optimum conditions, gasohol use may save nearly one gallon of gasoline and natural gas equivalent for every gallon of ethanol used.

o   HOW MUCH GASOHOL CAN BE PRODUCED?

In the 1980's there is the physical - though not necessarily economic - possibility of producing at least 5-10 billion gallons of ethanol per year, mostly from increased crops devoted to ethanol production. (This corresponds to 325,000-650,000 bbl./day or 4.5-9% of the current gasoline consumption of 110 billion gallons per year.) In the 1990's, however, the available land for energy crops could drop to a point where only about one third this amount could be produced. In addition, using conversion technologies currently under development, there is the physical possibility of producing 5 billion gallons of ethanol per year from crop residues, 10-20 billion gallons per year from increased forage grass production, and considerably more from wood. Due to a variety of factors, however, gasohol's practical potential will undoubtedly be considerably less than that which is physically possible.

Gasohol's practical potential depends on the time frame. In the next 3-5 years, domestic gasohol production will be limited primarily by the rate that new distilleries are built and idle capacity converted. Conversion should bring the fuel ethanol capacity to an estimated 40-90 million gallons per year by the end of 1980. This would yield 400-900 million gallons of gasohol per year compared to the current levels of 150-200 million gallons.

If gasohol is produced using coal or other non-premium fuels to supply energy for the distillation plant and marketed as a "regular"-grade transportation fuel, currently planned ethanol capacity could save 35-80 million gallons of gasoline and natural gas energy equivalent per year by the end of 1980 (2,300-5,200 bbl./day or 0.03-0.07% of current gasoline consumption.)

In addition to domestic production, there are plans to import 120 million gallons of ethanol per year from Brazil which would displace about 95 million gallons per year (6,000 bbl./day or 0.09% of current consumption) of gasoline. These imports would also increase the annual U.S. trade deficit by at least $50 million.

Because there is a 2 year lead time for distillery construction and start-up, the capacity that will come on line in 1981 depends on the current rate of investment in new distilleries. Although the available information is incomplete, there are at least 50-70 million gallons per year of new capacity which are under study or have been ordered.

Within the next decade, gasohol production could be limited by feedstock supply.

The longer term future of gasohol is still less assured. Future production costs are highly uncertain, due to uncertainties in future farm commodity prices, feedstock availability, and the cost of conversion processes using alternative feedstocks such as crop residues, grasses, wood and municipal solid waste. In addition, the development of less expensive octane boosters, or engine improvements which reduce the need for high

octane fuels, could jeopardize the utility of ethanol as an octane booster. This would alter the economics of fuel ethanol use and could reduce the demand.

These and other uncertainties may limit investment in ethanol distilleries to a total production level below that which is physically possible and can be sold profitably in the 1980's. It is equally possible that federal and state incentives may encourage the development of a large scale ethanol industry whose output may be difficult to market in the 1990's.

For these reasons, both the level of fuel ethanol production that will be achieved and the long term stability of price and demand are highly uncertain.

o   WILL GASOHOL PRODUCTION COMPETE WITH FOOD AND FEED PRODUCTION?

There are numerous sources of ethanol feedstocks, including food processing wastes, spoiled grain, and various substitutions among agricultural products which can free land for energy crop production. 1-2 billion gallons of ethanol per year (1-2% of current gasoline consumption) can probably be produced without a significant impact on food and feed prices. Beyond this ethanol production level, new cropland would have to be brought into production, and the farm commodity prices necessary to induce this land conversion are highly uncertain. Consequently, ethanol production levels significantly larger than 1-2 billion gallons per year if derived from food cropland, could lead to strong inflationary trends in food and feed markets, which would be a substantial indirect cost of ethanol production.

In the 1990's however, ethanol may be able to be produced competitively from cellulosic feedstocks (e.g., crop residues and wood), which would have little impact on food prices.

o   WHAT ARE THE COSTS?

Depending on the method of financing, distilleries should be able to sell ethanol (from $2.50/bu. corn) at between $0.91 and $1.11 per gallon plus delivery (currently $.10 to $.30 per gallon for stations outside the distillery's immediate locale). Due to the strong demand created by federal and state subsidies (totalling $0.40-$1.10 per gallon or $16.80-$46.20 per bbl. of ethanol) and intangible factors, ethanol was being sold for as much as $1.70 per gallon (F.O.B. the distillery) in June and July, 1979.

Gasohol would be competitive with gasoline costing the service station owner $.70/gallon (i.e., retail gasoline at about $0.99/gallon) if the ethanol costs $0.90-$1.00/gallon. With only the federal subsidy ($.04/gallon of gasohol or $16.80/bbl. of ethanol), gasohol can now compete

$1.30-$1.40 per gallon.

The current federal subsidy is adequate (although marginally so in some cases) to allow gasohol to compete with gasoline at today's ethanol production costs and gasoline prices.

It should be noted that current gasohol subsidies apply to imported as well as domestic ethanol, and the state plus federal subsidies on the planned imports of ethanol (120 million gallons per year) would amount to $50-$130 million per year.

o   CAN FARMERS PRODUCE THEIR OWN FUEL ON-FARM?

It is technically quite simple to produce ethanol containing 5% or more water on-farm. This alcohol could be used as a supplement to diesel fuel in retrofitted diesel engines, but even under favorable conditions the cost of the ethanol is about twice the current cost of the diesel fuel it would displace. Cost estimates significantly below this have been popular, but are based on questionable assumptions.

With slightly more sophisticated equipment, dry ethanol suitable for use in gasohol could be produced. On-farm production of dry ethanol could become competitive with commercially distilled ethanol if relatively automatic and inexpensive mass produced distilleries were available and if farmers charge little for their labor. As an economically profitable venture, however, on-farm ethanol production is, at best, marginal under present conditions.

For some farmers the cost and/or labor required to produce dry or wet ethanol may be of secondary importance. The value of some degree of energy self-sufficiency and the ability to divert limited quantities of corn and other grains when the price is low may outweigh the inconvenience and cost. As evidence, the Bureau of Alcohol, Tobacco, and Firearms expects to receive over 5000 applications for on-farm distillation permits this year.

o   WHAT ARE THE ENVIRONMENTAL IMPACTS?

All components of a gasohol "fuel-cycle" - growing and harvesting the biomass feedstock, converting it to alcohol, and using the gasoline/alcohol blend in automobiles - have significant environmental effects.

The choice of ethanol feedstock is the most critical factor determining the environmental impact of gasohol. As ethanol production grows beyond the feedstock capacity of surplus and waste materials, new land may be placed into intensive cultivation to provide additional feedstocks. If corn is the primary gasohol feedstock, the result will be a substantial increase in soil loss as well as fertilizer and pesticide use as millions of additional acres are put into production. The choice of other feedstocks will drastically alter these impacts; for example, using perennial grasses would considerably decrease erosion damages.

The major impact of alcohol distilleries -- potential degradation of water quality from the waste "stillage" -- can be prevented by byproduct recovery or waste treatment.

Despite claims of strong air quality benefits, gasohol use in automobiles appears to have a very mixed effect on automotive pollution. It is difficult to categorize the overall effect as either positive or negative, although carefully programmed use in selected locations (e.g., areas with high carbon monoxide concentration but no smog) or in selected segments of the automobile fleet (e.g., very richly tuned fleets in areas where hydrocarbons are more of a problem than oxides of nitrogen) may have unambiguously positive results.

o WHAT ARE THE SOCIAL IMPACTS?

The predominant social and economic impacts of gasohol production are the potential for new on-farm and other rural employment opportunities and the possibilities of conflicts between food and energy uses of cropland. The pace of development and the quantity of ethanol produced will be critical determinants of the social impacts. If the demand for fuel ethanol increases beyond the supply of feedstocks, competition between energy and food uses of land could result in more rapidly rising food prices and, eventually, more rapidly rising land prices. This would benefit landowners, but would hurt farmers who rent their land or who want to expand their holdings. Low and middle income groups would bear the greatest share of these costs. Although farming groups have supported gasohol initiatives in the hope that increased demand for corn would raise prices, historic experience indicates that rising land prices would absorb much of the profit.

If the demand for gasohol rises gradually and market imbalances are avoided, the overall social and economic impacts of fuel ethanol production could be strongly positive. On-farm and distillery employment could stabilize those rural communities which are currently experiencing unemployment problems.

INTRODUCTION

The development of near- to mid-term strategies for reducing our dependence on imported oil and natural gas is a major energy need of the United States. Displacing such imports with renewable domestic energy sources is a strategy that has generated significant popular and legislative interest. Gasohol in particular has become a focus of national attention, and it is to this subject that this report is addressed.

Gasohol is a mixture of one part ethyl alcohol (ethanol) and nine parts unleaded gasoline. Although automobiles could be designed to operate on alcohol alone, for the forseeable future the most economic use of ethanol is as an octane booster in gasoline. And of the synthetic liquid fuels from biomass, only ethanol can be produced with technology that is commercially available in the United States.

Although ethanol cannot by itself solve our energy problems, it may contribute to what must be a combination of national energy strategies. This report will place this contribution in perspective and clarify the role that gasohol may be expected to play in our energy future. Methanol has not been included because the technology for producing it from biomass is not commercially available at present and an adequate consideration of the resource base for methanol production would greatly expand the scope and complexity of the report. Methanol fuel, however, will be included in a later report on Energy from Biological Processes.

In this report, references are given as numbers in parentheses, with a full reference list at the end of the report. With the exception of the contractor report on Federal Bioenergy Programs, the various OTA contractor reports cited in the reference list will not be available for public distribution until the final report on Energy from Biological Processes has been released.

I. TECHNICAL ASPECTS OF GASOHOL

ETHANOL PRODUCTION

Commercial Distillation

The production of gasohol requires the integration of a number of factors, two of which are considered in this section -- ethanol distillation facilities and their feedstocks. Although ethanol can be produced from any feedstock capable of being reduced to the proper sugars, present U.S. production technologies rely on sugar and starch feedstocks. Suitable ethanol crops include corn, wheat, grain sorghum, sugarcane, sugarbeets, sweet sorghum, and Jerusalem artichokes. There is, however, no "best" ethanol crop, since different crops will be superior for ethanol production in different soil types and regions of the country. Current research into sugar and starch feedstock alternatives to corn are likely to produce strains that will outperform corn under some circumstances.

In addition to primary crop production, there are numerous sources of spoiled grain and food processing wastes that can be used to produce ethanol, but their total potential is small (1,2). Other processes are under development that will permit the commercial distillation of ethanol from cellulosic (cellulose containing) feedstocks such as crop residues, grasses, wood, and the paper contained in municipal solid waste.

The ethanol conversion process consists of four basic steps. First, the feedstock is treated to produce a sugar solution. The sugar is then converted in a separate step to ethanol and carbon dioxide by yeast or bacteria in a process called fermentation. The ethanol is removed by a distillation process which yields a solution of ethanol and water that cannot exceed 95.6% ethanol (at normal pressures) due to the physical properties of the ethanol-water mixture. In the final step the water is removed to produce dry ethanol. This is accomplished by adding to the solution a chemical that changes these physical properties and by distilling once again.

The material remaining in the water solution after the ethanol is distilled away, called "stillage," contains some dead yeast or bacteria and the material in the feedstock which was not starch or sugar. Grain feedstocks, for example, produce a high protein stillage (called distillers' grain) which can be used as an animal feed* while sugar and cellulose feedstocks produce a stillage with little protein and less feed value.

At the present time 15-20 million gallons of fuel ethanol per year are being produced commercially in the U.S., and domestic capacity should increase to 40-90 million gallons per year by the end of 1980.(3) New distilleries (e.g., 50 million gallons per year) can be brought on stream in only two years, and idle capacity can often be converted in one year or less. Beyond 1980 the information is sketchy, but there is at least 50-70 million gallons per year of new capacity which is under study or has been ordered and which can be in production in 1981.(3) In addition to domestic production, American Gasohol (Mineola, N.Y.) is planning to import 120 million gallons of ethanol per year from Brazil.(4)

------

* The exact nutritional value of distillers' grain in its various forms is still uncertain (e.g., the amounts that can be used in animal feed and the effect of using wet stillage as a feed). This is a subject that is currently being researched.

Looking towards the future, there are several processes under development which will be able to use cellulosic feedstocks as sources of sugar for ethanol production. All of the processes require higher capital investments (2-3 times higher than conventional processes)(5) because of expensive pretreatments, and this limits their present applicability. There are, however, a number of approaches to cellulose conversion which can improve its competitiveness by lowering the production costs of ethanol. These include lower capital charges through favorable financing, substantial credits for byproduct chemicals, improved ethanol yields (gallons per ton of feedstock), and process innovations. A process developed by Gulf Oil Chemicals Co., for example, uses municipal waste paper, and municipal bond financing would make the distillery competitive with conventional processes. (Because of the higher capital investment, special financing lowers the ethanol cost more than for grain distilleries.) Another cellulose process, developed in the U.S. during World War II and used commercially in the USSR, produces ethanol from wood. The process has recently been reevaluated(6) as a source of ethanol and byproduct chemical feedstocks (e.g., phenol and furfural). Although the capital investment for the distillery is three times that of a corn distillery, the byproduct credits are of sufficient value to make the ethanol competitive. The chemical industry, however, is unlikely to make the commitment to these feedstocks that would be necessary to support a large ethanol program until more information is available on the relative merits of biomass and coal derived chemical feedstocks. (7)

Aside from special financing or large byproduct credits, the key to making cellulosic feedstocks competitive is to achieve high ethanol yields without the use of expensive equipment, excessive loss of process chemicals, or the production of toxic wastes. At present there are no processes which fulfill all of these criteria. Current research and development efforts, however, could lead to significant results in 3-5 years, and commercial facilities could be available by the late 1980's.

Another way to reduce the costs of ethanol from cellulose is through process innovation. There are several possibilities for improvements, including minor changes which take advantage of the low purity requirements for fuel ethanol, and major process innovations for concentrating and drying the ethanol. An additional possibility involves a fundamentally different process for ethanol production. Rapidly heating cellulosic materials produces a gas which contains ethylene, a chemical that can be converted to ethanol with commercially available technology. Although this process is still at the laboratory stage, preliminary calculations indicate that the costs and yields could compare favorably to fermentation, and the process would require less energy. (8) It is unlikely, however, that the entire process could be made commercially available before the 1990's.

On-Farm Distillation

Apart from commercial distilleries, there has been interest expressed in the role which individual farmers can play in ethanol production. Producing ethanol on the farm, however, faces a number of limitations which may severely restrict its widespread practice.

On site distillation of ethanol for farm use may be possible at a cost of $1.00/gallon of 95% ethanol plus labor.* If it is used as a fuel supplement for retrofitted diesel driven tractors this would be equivalent to diesel fuel costing $1.70 per gallon. (9) Current diesel prices would therefore have to double for ethanol production to be competitive without subsidies.

Cost estimates significantly below this are apparently based on ignoring equipment and/or fuel costs, assigning a credit for the byproduct animal feed that is significantly higher than its market value, and/or producing ethanol containing significantly more water than the 5% assumed above.

If the purpose of on-farm distillation is to develop a degree of energy self-sufficiency, the higher cost of ethanol may be acceptable. Due to technical limitations, however, ethanol can displace only 35% of the diesel fuel used in retrofitted diesel engines. (9)

Limitations also apply where distillation is viewed as a process for diverting significant quantities of grain produced on the farm. A typical farm of 500 acres could produce 50,000 bu. of corn, of which 1,000 bu., or 2%, would provide as much ethanol (2,500 gallons) as could reasonably be

---

* While equipment costs will vary considerably depending upon how automatic they are, $1 for each gallon per year of capacity is plausible. Assuming this equipment cost, the costs per gallon of ethanol are: $0.58 for net feedstock cost, $0.20 for equipment costs (operated at 75% of capacity), $0.20 for fuel (assuming $3/MMBTU and 67,000 BTU/gallon) and $0.05 for enzymes and chemicals, resulting in $1.03/gallon of ethanol or $0.98/gallon of 95% ethanol.

used as a diesel fuel substitute in retrofitted diesel engines. (9, 10) Converting 20% of the crop to ethanol would produce 25,000 gallons, far more than could be used on the farm, and would require a significant investment of probably $25,000 or more.

The quality of the ethanol most easily produced on farms across the nation is likely to limit the uses for which it would be appropriate. As a gasoline additive, for example, ethanol must be free of water in order to avoid operating problems. (9, 11) Not only would producing dry ethanol change the economics of on-farm distillation, but the current drying processes involve the use of dangerous chemicals. Alternate processes using drying agents, or desiccants, can probably be developed, but the costs are uncertain. (12)

On-farm production of dry ethanol could become competitive with commercially distilled ethanol if relatively automatic mass-produced distilleries could be sold for less than $1 for each gallon per year of capacity and if farmers charge little for their labor. Although 150,000 gallon per year package distilleries producing 95% ethanol can be bought for prices approaching this value (13), OTA is not aware of any farm size (1,000 - 10,000 gallon per year) package distilleries for producing either 95% or dry ethanol. The price goal for automatic, on-farm, dry ethanol production is not unrealistic, but it will probably require process innovations, particularly in the drying step, and could well involve the use of small, inexpensive computers for monitoring the process.

Other concerns about on-farm distilleries involve the fuel used to operate them and the possible diversion of the alcohol produced. As with large distilleries, abundant or renewable domestic energy sources should be used to fuel the on-farm distilleries, the most appropriate of which may be crop residues (using gasifiers currently under development). Another possibility is the use of solar powered distilleries, but the costs are uncertain.* Obtaining appropriate energy sources for distilleries need not be a problem, but it is something that should be considered in legislation designed to encourage on-farm distillation.

The ethanol produced with most processes can easily be converted to beverage alcohol.** Although ethanol can be denatured to render it unfit for consumption, regulations will be difficult to administer.

For some farmers the cost and/or labor required to produce ethanol may be of secondary importance. The value of some degree of energy self-sufficiency and the ability to divert limited amounts of corn and other grains when the market price is low may outweigh the inconvenience and/or costs. The Bureau of Alcohol, Tobacco and Firearms has received over 2,800 applications for on-farm distillation permits and they expect 5,000 by the end of the year. (14) As a profitable venture, however, on-farm production of ethanol is, at best, marginal.

---

* The key problem is that concentrating a 10% ethanol solution to 95% ethanol requires a theoretical minimum of 25 evaporation-condensation cycles. Since a solar powered water distillery only produces one evaporation-condensation cycle, the ethanol solution would have to be put through a still of this type many times in order to concentrate the ethanol. Designs better suited to ethanol concentration, however, can probably be developed.

** Dilute to 50% with water and filter through activated charcoal.

GASOHOL AS AN AUTOMOBILE FUEL

Between 150-200 million gallons of gasohol per year are being sold in over 800 service stations in at least 28 states, and the major U.S. automobile manufacturers have extended their warrantees to permit the use of gasohol. (2, 15) Despite its increasing acceptability and use, however, there are several technical aspects which merit consideration. These include fuel stability, drivability, octane boosting properties of ethanol, and mileage with gasohol. These points are considered below.

Fuel Stability

Only minute quantities of water will dissolve in gasoline and although the addition of ethanol increases the water solubility somewhat, gasohol containing more than 0.3% water can separate into two phases, or layers, which can cause automobiles to stall. Although some additives designed to prevent phase separation with 95% ethanol have been tested, none has proven to be fully satisfactory. (9) Consequently, gasohol blends require dry (anhydrous) ethanol, and storage and transport tanks must be kept free of moisture. Although there have been occurrances of phase separation in a few service stations, if dry ethanol is used and due care taken, this should not be a significant problem.

Automobile Performance

For most automobiles, performance with gasohol should be indistinguishable from that with a gasoline of the same octane. A small but unknown fraction of the existing automobile fleet, however, will experience surging, hesitation, and/or stalling with gasohol, due to a variety of causes.* (9) But these problems should disappear with time as gasohol use

---

* The leaning effect of gasohol, damage to gaskets, pump diaphragms, etc., and dislodging of deposits in the fuel system leading to clogging in the fuel filter and/or carburetor.

becomes more widespread and the automobile fleet is replaced with new cars manufactured to accept gasohol.

Octane

An important advantage of gasohol is that its octane* is higher than the gasoline to which the ethanol has been added. The exact increase will depend on the octane and composition of the gasoline and can vary from an increase of 0.8 to 5 or more octane numbers. (16) For "average" gasolines the increase is about 3-4 octane numbers.

Raising the octane of motor fuels would enable automobile manufacturers to increase the efficiency of automobile engines, but this is unlikely to occur unless gasohol is widely available. Alternatively, the octane of the gasoline blended to gasohol can be lowered to exactly compensate for the octane boosting properties of the ethanol. If this is done, there is an energy savings at the refinery of 88,000-150,000 BTU per barrel of oil refined. (9, 16, 17) If these energy savings are attributed solely to the ethanol, a savings of 0.27-0.45 gallons of gasoline equivalent can be achieved for each gallon of ethanol used.** Achieving this savings, however, will require the cooperation of oil refiners and distributors.

----

* Average of motor and research octane.

** The median energy savings is 118,000 BTU per barrel of crude oil refined (higher heat content). Since an average of 55% of a barrel of crude oil is turned into gasoline, and this gasoline is mixed with one ninth as much ethanol, then the 118,000 BTU savings is attributed to about 2.6 gallons of ethanol. With a higher heat content of 125,000 BTU/gallon for gasoline, this results in 0.36 gallons of gasoline per gallon of ethanol.

In order to realize the energy savings of ethanol it is essential that car owners use a fuel with the correct octane. If drivers buy a higher octane gasohol than their cars require, based perhaps on advertising claims of its superiority, the energy savings would be negated.

Mileage

Ethanol contains less energy per gallon than gasoline, and a gallon of gasohol contains 3.8%* less energy than a gallon of gasoline. If all other factors were equal, this would result in 3.8% lower mileage (miles per gallon). The gasohol, however, also "leans" the fuel mixture (i.e., moves the air-fuel mixture to an effective value that contains less fuel and more air) which increases the thermal efficiency (miles per BTU) in many cars, but lowers it in a few.

The mileage measured for gasohol varies considerably from test to test, but road tests have often registered better mileage averages than laboratory tests. The results of the road tests, however, are less accurate than

---

* The lower heat content of ethanol and gasoline are about 76,000 BTU/gallon and 117,000 BTU/gallon, respectively. In addition 0.9 gallon of gasoline plus 0.1 gallon of ethanol results in 1.002 gallons of gasohol. Blending the alcohol result in a 3.6% drop and the expansion an additional 0.2%. This number, however, can vary somewhat for different gasoline compositions.

laboratory tests* and have sometimes been conducted on vehicles which are not representative of the U.S automobile fleet. Based on laboratory data, the mileage (miles per gallon) for gasohol is expected to average 0-4% less than for gasoline.

ENERGY BALANCE

The energy objective of an ethanol fuel program is the displacement of foreign oil and gas with a domestic synthetic fuel. The impact of such a program depends upon the energy balance of growing the feedstock, converting it to ethanol, and using the ethanol as fuel. The fuels used in the conversion process must also be considered.

For each gallon of ethanol derived from corn, farming and grain drying consume, on the average, the energy equivalent of 0.29 gallons of gasoline

---

* The data available from the 2 million mile gasohol test, (18) for example, have been analyzed by OTA. Using a standard statistical test ("t" test) reveals that the spread in data points (standard deviation) is so large that the mileage difference between gasohol and regular unleaded would have to be more than 30% (2 times the standard deviation) before OTA would consider that the test had demonstrated a difference in mileage. While more sophisticated statistical tests might indicate that the measured difference in mileage is meaningful, the validity of these statistical methods is predicated on all the errors being strictly random; and the assumption of random errors is suspect unless the number of vehicles in the test fleet is orders of magnitude larger than any tests conducted to date.

in the form of oil (for fuel and petrochemicals) and natural gas (for nitrogen fertilizers). (10) The exact amount, however, will vary with farming practices (e.g., irrigation) and yields. Although corn is often cited as an energy intensive crop (due to the high energy inputs per acre cultivated), the energy used per ton of corn grain produced is comparable to results achieved with other grains. (10) In general, however, the energy input per gallon of ethanol produced will increase when the farmland is of poorer quality (e.g., set-aside acreage) and/or in dryer or colder climates (i.e., most of the western half of the country, excluding Hawaii).

The fuel used in the distillation process is perhaps the most important factor in determining the displacement potential of ethanol. Even under the most favorable circumstances, distillery energy consumption is significant. Although the distillery producing most of the fuel ethanol used today reportedly consumes 0.25 gallon of gasoline equivalent (0.24 in the form of natural gas) per gallon of ethanol, (14) the derivation of this number involves some arbitrary decisions about what energy inputs should be attributed to the facility's food processing operations, and various factors probably would make the process unsuitable* for extensive use in fuel ethanol production. Energy

---

* 1) The distillery probably uses waste heat from an adjacent byproduct processing plant which consumes nearly as much energy as the distillery but is not included in the energy balance cited here, 2) acetaldehyde is left in the ethanol which increases evaporative emissions and the possibility of vapor lock in automobiles, and 3) the economics are predicated on credits for byproducts (e.g., corn oil), whose markets could be saturated.

efficient stand alone fuel ethanol distilleries would consume the equivalent of 0.4-0.6 gallon of gasoline per gallon of ethanol. (5, 20) To maximize the displacement potential of ethanol it is therefore essential that distilleries use abundant or renewable domestic energy sources such as coal, biomass, and/or solar heat. As shown in the following table, reliance on these fuels would reduce the total use of oil and gas at the distillery to insignificant levels.

The amount of petroleum displaced by ethanol fuel also depends on the manner in which it is used. As an additive in gasohol, each gallon of ethanol displaces about 0.8 gallons of gasoline. If the oil refinery produces a lower grade of gasoline to take advantage of the octane boosting properties of ethanol, an additional 0.36 gallon of gasoline equivalent can be saved in refinery processing energy, as described on page 11.

Additional energy savings are achieved by using the byproduct distillers' grain as an animal feed. To the extent that crop production is displaced by this animal feed substitute, the farming energy required to grow the feed crop is displaced.

In all, the total displacement of imported fuels achieved per gallon of ethanol can be increased by a factor of 2.5 by requiring that 1) petroleum and natural gas not be used to fuel ethanol distilleries and 2) lower octane gasoline be used in gasohol blends.

Table 1 summarizes the oil and natural gas used and displaced for the entire gasohol fuel cycle. The quantities are expressed as gallons of gasoline equivalent for each gallon of ethanol produced and used

Table 1 Energy Balance of Gasohol from Corn

Oil and Natural Gas Used (+) and Displaced (-)

(in gallons of gasoline equivalent per

gallon of ethanol produced and used[a])

| | Present | Set Aside and Potential Cropland | |
		Coal Fired Distillery	Coal Fired Distillery & Lowering of Gasoline Octane
Farming	0.29	0.35[b]	0.35[b]
Distillery	0.24	0[c]	0[c]
Distillery Byproduct	-0.09[d]	-0.09[d]	-0.09[d]
Automobile	-0.80	-0.80	-0.80
Oil Refinery	-	-	-0.36
Total	$-0.36(\pm 0.3)$	$-0.54(\pm 0.3)$	$-0.90(\pm 0.3)$

a   Lower heat content of gasoline and ethanol taken to be 117,000 BTU/gallon and 76,000 BTU/gallon, respectively.

b   Estimated uncertainity of ± 0.15.

c   50,000-70,000 BTU of coal per gallon of ethanol.

d   Assumed that distillers' grain replaces corn grown on average cropland.

Source: OTA

in gasohol. The three cases presented correspond to (1) the present situation, (2) future production of ethanol from the less productive land that can be brought into crop production and using coal as a distillery fuel, and (3) the same as (2) except that the octane of the gasoline is lowered to exactly compensate for ethanol's octane boosting properties. These cases result in net displacements of (1) slightly more than 1/3 gallon, (2) slightly more than 1/2 gallon and (3) slightly less than 1 gallon of gasoline and natural gas equivalent per gallon of ethanol used.

It should be noted, however, that if oil or natural gas is used as a distillery boiler fuel, the second case could result in the fuel cycle consuming slightly more oil and natural gas than is displaced. This is the situation that is alluded to in most debates over gasohol's energy balance, but it is a situation that can be avoided with appropriate legislation.

In the most favorable case (case (3) above) and with an energy efficient distillery, however, the ratio of total energy displaced to total energy consumed is 1.5 ($\pm$0.4), i.e., the energy balance is positive (a ratio greater than 1). And if the feedstocks are derived from more productive farmland, or local conditions allow energy savings at the distillery (e.g., not having to dry the distillers' grain), then the balance is even more favorable. Alternatively, an energy credit could be taken for the crop residues, which would also improve the calculated balance. This general approach to the energy balance, however, does not consider the different values of liquid versus solid fuels.

The uncertainty factor in Table 1 of plus or minus 0.3 gallons of gasoline per gallon of ethanol is due primarily to inherent errors in fuel efficiency measurements, differences in farming practices and yields, and the magnifying effect on these errors of the low (10%) ethanol content of gasohol. These factors make it unlikely that more precise estimates can be made in the near-term.

II.   GASOHOL ECONOMICS

ETHANOL COSTS

Ethanol costs* are influenced by the capital investment in and financing of the distillery, the distillery operating costs, and the byproduct credits. The cost of an ethanol distillery for starch and sugar feedstocks is about $1.00-$2.00 for each gallon per year of capacity. Distilleries that rely upon sugar feedstocks are more expensive than those using starch due to the equipment needed to handle the feedstock and to concentrate the sugar solution to a syrup for storage. Coal-fired distilleries are more expensive than oil or gas fueled distilleries, due to the costs of coal handling and pollution control equipment.

For a coal-fired 50 million gallon per year distillery using starch feedstock, the capital related charges are about $0.35-$0.45 per gallon of ethanol, assuming 100% private equity financing and a 13% after tax return on investment. The comparable figure for 100% debt financing with favorable terms is $0.15-$0.25 per gallon. These charges, however, can vary significantly with depreciation allowances, tax credits and other economic incentives.

The major operating expenses are the fuel and feedstock costs. The coal ($30/ton) would cost about $0.10/gallon of ethanol, which is sufficiently less than oil or natural gas to compensate for the added costs of the coal boiler and handling and pollution control equipment. Although increased demand could raise coal prices, the effect on the ethanol costs would be relatively small.

---

* All dollar figures quoted here are for 1978 and are in 1978 dollars.

The largest cost in ethanol production is the net feedstock cost, or the feedstock cost less the byproduct credit. With corn at $2.50 per bushel, the corn grain costs $0.96 per gallon of ethanol and the byproduct credit is about $0.38 per gallon, resulting in a net feedstock cost of $0.58 per gallon. Since farm commodity prices are extremely volatile, the net feedstock and resultant ethanol cost are also variable. A $0.50/bu. increase in corn grain prices (and a proportionate increase in the byproduct credit), for example, would raise the ethanol cost by $0.12 per gallon.

Distilleries which rely on grain feedstocks depend for their byproduct credit on the cost of distillers' grain as an animal feed supplement. There is uncertainty, however, regarding the amounts of distillers' grain which can profitably be added to animal feeds. USDA and others have estimated that byproduct credits could begin to drop due to saturation of the domestic feed market at about 2 billion gallons of ethanol production per year (0.13 million bbl./day of ethanol or about 1.8% of the present gasoline consumption). (10, 21, 22) At significantly higher levels of production, new markets for distillers' grain (e.g., exports, protein extracts) would have to be developed or distillers could lose the byproduct credit, increasing the ethanol cost by $0.38 per gallon.

The costs for ethanol produced from various feedstocks are shown in Tables 2 and 3. Although the costs will vary depending on the size of the distillery, ethanol can be produced from corn ($2.50/bu.) in a coal fired 50 million gallon per year distillery for $1.11 (+$0.10) per gallon with 100% private equity financing (including a 13% return on investment) and $0.91 (+$0.10) per gallon with 100% debt financing.* About $0.10-$0.30 per gallon

---

* Details are given in note d of Table 2.

TABLE 2

Late 1978 Production Costs for Ethanol
From Grain and Sugar Crops
In a 50 Million Gallon Per Year Distillery

	Grain[a]	Sugar[b]
Fixed Capital	$59 million	$100 million
Working Capital (10% of F.C.)	$5.9 million	$10 million
Total Investment	$64.9 million	$110 million

Operating Costs:

	$/gallon of 99.6% ethanol	
Labor	0.04	0.05
Chemicals	0.01	0.01
Water	0.01	0.01
Coal ($30/ton)	0.09	0.00[c]
Sub total	0.15	0.07

Capital Charges:

15%-30% of Total Investment per year[d]	0.18 - 0.38	0.33 - 0.66
Total	0.33 - 0.53	0.40 - 0.73

a) Includes drying of distillers' grain

b) Includes equipment for extracting the sugar from the feedstock concentrating it to a syrup for storage.

c) Bagasse fueled distillery appropriate for sweet sorghum and sugarcane.

d) There are many, often complex formulae to compute actual capital costs. Economic factors considered include debt/equity ratio, depreciation schedule, income tax credit, rate of inflation, terms of debt repayment, operating capital requirements, and investment lifetime. However, a realistic range of possibilities for annual capital costs would lie between 15% and 30% of total capital investment.

The upper extreme of 30% may be obtained assuming 100% equity finance and a 13% after tax rate of return on investment. The lower extreme of 15% may be obtained assuming 100% debt financing at a 9% rate of interest. Both calculations assume constant dollars, a 20 year project lifetime, and include a charge for local taxes and insurance equal to 3% of fixed capital costs. For a more detailed treatment of capital costs see OTA, Application of Solar Technology to Today's Energy Needs, Vol. II, Chapter 1.

Source: OTA and Reference 20.

TABLE 3

Cost of Ethanol From Various Sources

Feedstock	Price[a]	Net Feedstock Cost[b] ($/gallon ethanol)	Ethanol Cost ($/gallon)	Yield[c] (gallons of ethanol per acre)
Corn	$2.44/bu	0.57	0.90-1.10	220
Wheat	$3.07-4.04/bu[d]	0.73-1.08[d]	1.06-1.61	85
Grain Sorghum	$2.23/bu	0.49	0.82-1.02	130
Oats	$1.42/bu	0.59	0.92-1.12	75
Sweet Sorghum	$15.00/ton[e]	0.79	1.19-1.52	380[e]
Sugar Cane	$17.03/ton[f]	1.26	1.66 - 1.99	520

———

a) Average of 1974-77 seasonal average prices.

b) The difference in feedstock costs might not hold over the longer term due to equilibration of prices through large scale ethanol production.

c) Average of 1974-1977 national average yields.

d) Range due to different prices for different types of wheat.

e) Assuming 20 fresh weight tons/acre yield, $300/acre production cost.

f) Excludes 1974 data due to the anomalously high sugar prices that year.

SOURCE: USDA, Agricultural Statistics, 1978 and OTA.

should be added to these costs for deliveries of up to 1,000 miles from the distillery. (The ethanol is currently delivered in tank trucks, but as the production volume grows other forms of transportation, such as barge shipments, rail tank cars, and petroleum product pipelines,[*] could lower the cost to as low as $0.03 - $0.05 per gallon under favorable circumstances.)

---

[*] Various strategies can be used to eliminate potential problems with the water sometimes found in petroleum pipelines. If ethanol is being transported, the total volume of ethanol in the batch can be kept large enough so that the percentage of water in the delivered ethanol is within tolerable limits. If gasohol is transported, it can be preceded by a few hundred bbl. of ethanol which will absorb any water found in the pipeline, thereby keeping the gasohol dry. Other strategies also exist or can be developed. (23)

VALUE OF ETHANOL IN GASOHOL

For the purpose of this report, value is defined as the price at which ethanol is competitive as a gasoline additive. Calculated simply on the basis of its energy content, ethanol costing $1.10/gal. is equivalent to gasoline selling at the refinery gate for $1.70/gal. (2.5 times the present price), or $44/bbl. crude oil.[*]

The value of ethanol in gasohol, however, is primarily determined by its octane boosting properties. Although this varies considerably depending on the gasoline and other specifics, OTA estimates the value at 1.9-2.5 times the average crude oil acquisition price (see box on page 26 for the details).

Without subsidies, ethanol presently (July, 1979) has a value of $0.75-$1.00 per gallon. With the federal subsidy of $0.40 per gallon of ethanol ($16.80/bbl. of ethanol or $0.04 per gallon of gasohol), the value is $1.15-$1.40 per gallon; and with some state subsidies of $0.40-$0.70 per gallon ($16.80-$29.40/bbl.) of ethanol, the value is $1.55-$2.10 per gallon.

Ethanol distilled from corn ($2.50/bu.) can be produced in a 50 million gallon per year coal fired distillery and delivered to a service station for $1.20-$1.40 per gallon, making it competitive with the federal subsidy alone if the gasohol is blended at the service station. At this price ethanol

---

[*] Assuming the current value of 1.64 for the ratio of the refinery gate price of unleaded regular to the crude oil acquisition price. (24)

would be competitive without subsidies when U.S. refineries pay an _average_ crude oil price of $20-$31/bbl., or when retail unleaded gasoline costs about $1.10-$1.60 per gallon* on the average.

Several factors, however, can change the estimated value of ethanol. If a special, low octane, low vapor pressure gasoline is sold for blending with ethanol, at low sales volumes the wholesaler might assign a larger overhead charge per gallon sold. Also, the refinery removes relatively inexpensive gasoline components in order to lower the vapor pressure** of the gasoline, and this increases its cost. On the other hand, in areas where gasohol is popular, the large sales volumes lower service station overhead per gallon of gasohol, thus raising ethanol's value. These factors can change the value of ethanol by as much as $.40 per gallon in either direction; and the pricing policies of oil refiners and distributors will, to a large extent, determine whether ethanol is economically attractive as an octane boosting additive.

---

* Assuming cost relationships, as follows: Refinery gate price equal to 1.64 times crude oil prices plus delivery and retail mark-ups and taxes totalling $0.30-$0.40/gallon. (23)

** The more volatile components of gasoline (e.g., butanes) may be removed to decrease evaporative emissions and reduce the possibility of vapor lock. Although these components can be used as fuel, removing them decreases the quantity of gasoline and the octane boost achieved by the ethanol. Consequently, the advantages of having a less volatile gasoline must be weighed against the resultant decrease in the gasoline quantity and the value of the ethanol. Further research is needed to help resolve the dilemma.

Two Ways to Calculate the Value of Ethanol

Two different values for ethanol can be derived, depending upon where the ethanol is blended to form gasohol.

At the oil refinery, each gallon of ethanol used as an octane booster saves the refinery the equivalent of 0.36 gallons of gasoline by allowing the production of a lower octane gasoline (see section on octane under Technical Aspects of Gasohol). In addition, the gallon of ethanol displaces 0.8 gallons of gasoline directly (2% mileage decrease with gasohol), leading to a total displacement of 1.16 gallons of gasoline per gallon of ethanol. At the refinery gate, unleaded regular costs about 1.64 times the crude oil price, so the ethanol is valued at 1.16 x 1.64 = 1.9 times the crude oil price.

If the gasoline retailer blends the gasohol, the value of the ethanol is somewhat different. Gasoline retailers buy regular unleaded gasoline for about $0.70 per gallon (24) and sell gasohol for a rough average of $0.03 per gallon more than regular unleaded. (9) (The difference between this and the retail price of gasoline is due to taxes and service station mark up, which total about $0.29/gallon. (24) One tenth gallon of ethanol displaces $0.07 worth of gasoline and the mixture sells for $0.03 per gallon more. Therefore, 0.1 gallon of ethanol is valued at $0.10 or $1.00/gallon. This is 2.5 times the July, 1979 average crude oil price of $0.40 per gallon.

Both of these estimates are approximate, and changing price relations between crude oil and gasoline can change the estimates.

SOURCES OF ETHANOL

In the course of developing a large-scale gasohol program, ethanol supplies could be increased by taking advantage of such sources, methods or strategies as the following:

- o spoiled and substandard grain

- o food processing wastes

- o direct imports of ethanol

- o reduction of grain exports

- o cultivation on set-aside and diverted cropland

- o substitution among crops

- o substitution of forage for ethanol feedstock crops in livestock rations

- o cellulose feedstocks after the late 1980's.

Spoiled and substandard grains and food processing wastes can be utilized to produce ethanol totaling somewhat less than 1% of current gasoline consumption.(1, 2) In some cases, however, they are an unreliable source of supply, or are locally available only in small quantities. Realizing their full production potential will probably involve using them as feedstock supplements for distilleries relying on other sources.

Ethanol can be imported from Brazil for prices lower than it is being produced domestically. Since the imported ethanol costs a minimum of $0.42 per gallon more than the crude oil it could displace, the planned level of imports (120 million gallons per year) would increase our trade deficit by at least $50 million,* and federal plus state subsidies totalling $50 million to $130 million would be paid in the process.

Policies intended to limit the export of grains, or policies which effectively reduce exports by deliberately raising the price of exported grains (e.g., by fixing the price of corn to that of crude oil), can result in additional feedstocks for ethanol production. Recent grain exports have been 70-80 million metric tons/year. These exports could produce about 6-9 billion gallons of ethanol per year, displacing approximately $3-5 billion in imported crude oil. The loss of $10-12 billion in grain export revenues, however, would result in a $5-9 billion net increase in the trade deficit.

With corn at $2.50 bu., imported crude oil would have to cost about $32-$40/bbl. before it would decrease the trade deficit to curtail corn exports to increase the supply of ethanol feedstocks.** When economic forces (e.g., rising prices) reduce the level of grain exports, however, the situation is more subtle. Increasing the prices of grain would decrease the volume of exports, but it might initially increase slightly the gross income from exports. As grain prices continued to rise, however, the gross income from exports would eventually drop.

---

* According to the importer, American Gasohol, the import price is at least $1.00/gallon. (4) Each gallon of ethanol, as it is used now, displaces less than 0.8 gallons of crude oil at $0.50/gallon ($21/bbl). If the octane boosting properties of ethanol are exploited, the displacement is less than 1.16 gallons of crude oil per gallon of ethanol. Therefore $1 worth of ethanol would displace less than $0.58 worth of crude oil, resulting in a net increase in the trade deficit of at least $50 million.

** The situation is more favorable if the distillers' grain byproduct can be exported instead of the corn. In this case, there would be no net change in the trade deficit with the current prices of corn and distillers' grain and with crude oil at $20-$25/bbl., which is near the current price. Pursuing this strategy, however, would increase the international price of corn and decrease the international price of distillers' grain. Consequently, crude oil prices would have to be somewhat higher than $20-$25/bbl. for the strategy to decrease the trade deficit.

Cultivation on set aside and diverted acreage is often mentioned as a possible source of ethanol feedstocks. In 1978 there were 18.2 million acres in these categories and the 1979 total is about 11.2 million acres.(25) Although the majority of this land is not suitable for corn production, sufficient feedstocks could have been produced in 1978 and 1979 for about two and one billion gallons of ethanol, respectively. The quantity of set-aside and diverted acreage, however, will vary significantly from year to year and there is no assurance that this land will continue to be available for energy production.

In addition to set-aside and diverted cropland, OTA estimates that at least 30 million acres of potential cropland and cropland pasture can be used for the production of ethanol feedstocks in the 1980's over and above the land required for food, feed, and fiber production. (26) This would be sufficient to produce 5-7 billion gallons of ethanol per year.

Crop yields for this land, however, are likely to be more sensitive to weather variations[*] than the land currently under cultivation. Consequently, a heavy reliance on this land for grain production is likely to increase the year to year variability in grain supplies. This could lead to greater fluctuations in farm commodity prices and could require a larger grain buffer stock to stabilize prices. The required size of the buffer stock, and its added costs, are unknown, but increasing the buffer stock by 10% of the additional grain produced would cost about $0.01 per gallon of ethanol in federal grain storage subsidies ($0.25/bushel year).

---

[*] An often cited reason that this land is not now in production is that the soil does not retain moisture well or is prone to periodic flooding. Consequently the crop growth could be very sensitive to the rainfall pattern and could vary significantly from year to year.

The cost of converting this land to crop production varies from negligible amounts to perhaps $600/acre for some forested land. (26) Although federal grants could eliminate the one time cost of conversion, it is not known how much land would actually be brought into production at any given level of farm commodity prices (see next section). Consequently, the full cost of utilizing this land is unknown.

As the demand for ethanol feedstocks increases and more distillers' grain becomes available several types of market induced substitutions can occur. The distillers' grain can substitute for soybean meal in animal feed, which could result in less soybean production. Land which is presently in soybeans could then be used for additional ethanol feedstock production. In addition, some feed corn could be replaced with a combination of forage grass and distillers' grain. There are numerous uncertainties, however, about how much substitution actually will take place[*] and how much distillers' grain can profitbly be fed to animals. Assuming these substitutions occur, the total quantity of ethanol could possibly be raised from the 5-7 billion gallons per year from potential cropland and cropland pasture to as much as 10 billion gallons per year.

In the 1990's, the quantity of land available for energy crop production beyond the needs for food, feed and fiber will probably drop and ethanol producers may have to convert to cellulosic feedstocks. The potential ethanol production from these sources[**] is over 5 billion gallons

---

[*] The soybean meal industry, for example, may continue to buy soybeans and attempt to export the meal. The major customer, however, would probably be the EEC, which might impose import restrictions in order to protect its indigenous soybean meal industry. As a result there could be severe competition between distillers' grain and soybean meal, and the outcome is uncertain.

[**] Assuming potential yields of 100 gallons of ethanol per ton of feedstock.

per year from crop residues, an additional 10-20 billion gallons per year from increased forage grass production, and considerably more from wood. And based on OTA's assessment of municipal solid waste, (27) an additional 3-4 billion gallons per year could be obtained from paper derived from this source.

With the potential availability of grain feedstocks, the production of ethanol in the next 3-5 years will be limited primarily by the rate that new distilleries are built. Although production could conceivably reach a level of 7-10 billion gallons per year by the 1990's, expanding the total capacity to a level above 1-2 billion gallons per year would make ethanol production compete increasingly with other uses for farm commodities. In the mid- to long-term this competition may be severe, and to maintain or expand a fuel ethanol industry, distilleries may have to turn to cellulosic materials for their feedstocks.

COMPETITION BETWEEN FOOD AND FUEL

At this early stage in the development of the ethanol fuel industry, the cost of feedstock is tied directly to the value of farm commodities as food. As the ethanol industry expands, however, this relationship could reverse itself. A combination of ethanol subsidies and rising crude oil prices could drive up the price of farm commodities and ultimately the price of food. The extent to which this will happen depends critically upon how much additional cropland can be brought into production in response to rising food prices and, eventually, on the cost of producing ethanol from cellulosic feedstocks. These and other major uncertainties, such as future weather and crop yields, make it impossible to predict the full economic impact of a large fuel ethanol program.

The relatively low demand for fuel ethanol feedstocks currently exerts negligible pressure on farm commodity prices. As long as fuel ethanol production is sufficiently profitable, however, new distilleries will be built and feedstock purchases will expand. The increased demand will drive corn prices up toward the distillery break even point and thereby increase the price for all purchasers of corn. Under these circumstances food consumers would be indirectly subsidizing the consumption of fuel.

This indirect subsidy is illustrated in the following example. If the price at which ethanol is competitive increases by $0.12 per gallon, due to increased subsidies or a $2.50/bbl. increase in crude oil prices, corn prices would eventually increase by $0.50/bu. Domestic consumption of 4 billion bushels of feed corn (1976-1977) would cost an additional $2 billion. Although there would be a number of market adjustments, the increased corn cost would eventually appear as higher prices for meat and other food products. Excluding downstream markups, U.S. food expenditures could increase by more than 1%. Farm income, however, could increase by more than 3.5%.

The cost of this indirect subsidy per gallon of ethanol would depend on the supply response to increased corn prices. If ethanol production increased 500 million gallons (about 25 times the current fuel ethanol production) in response to a $0.12 per gallon increase in the price at which ethanol is competitive, the indirect subsidy would still be more than $4 per gallon of ethanol. If the supply response were ten times larger, 5 billion gallons, the indirect subsidy could be more than $0.40/gallon.

The previous example is perhaps an oversimplification -- actual impacts on feedstock prices and consumer food expenditures may be larger or smaller, depending on a complex of economic factors. Economic forces, however, will tend to couple the prices of food and fuel and transfer instabilities from one sector to the other. Although rising fuel prices will increase farm commodity prices in any case, a large fuel ethanol program could involve significant indirect costs and increase the inflationary impact of rising fuel prices, unless the program is designed to restrain the purchase of ethanol feedstocks in times of short supply. This would of course greatly increase the financial risks for ethanol producers and make the supply of ethanol uncertain.

COMPETITION WITH OTHER LIQUID FUELS

Whether or not ethanol is worth its cost, including both direct and indirect subsidies depends upon the cost and availability of other liquid fuels and the cost of energy conservation. Ethanol shares an advantage with existing conservation technologies in that it uses current technology and thus it may be an important fuel during the 1980's before possibly less expensive synfuels and newer or improved conservation technologies can be made available. Table 4 permits cost comparisons among some alternative fuel sources.

As an octane boosting additive, ethanol is nearly competitive today. The development of less expensive octane boosters or automobile engines which do not require high octane fuels, however, could seriously curtail the market for ethanol as an octane booster. In this case, ethanol would have to be marketed on its fuel value alone.

As a stand alone fuel, ethanol is unlikely to be competitive with methanol from coal, but it might be competitive as a fuel additive to the more expensive synfuels. The cost uncertainties, however, are too great to reliably predict whether a strong demand for fuel ethanol will continue into the 1990's.

The long-term viability of the fuel ethanol industry, will depend not only on sustained market demand, but also on the costs of conversion processes utilizing cellulosic feestocks. A major constraint may be the availability of capital for the large investments that are likely to be needed to convert distilleries to the cellulosic processes. These investments, for example, could be as large or larger than the cost of new

TABLE 4

ESTIMATED COSTS IN 1978 DOLLARS OF ALTERNATIVE LIQUID FUELS[1]

Fuel Source	$/MMBTU (Raw Liquid)	$/MMBTU (Refined Motor Fuel[2])	1990 Potential (000 bbl./day)
Fuels Requiring No Automobile Modification			
Imported Crude	3.40	6.20	4500 - 8500
Enhanced Oil Recovery	1.70 - 5.90	3.10 - 10.90	300 - 1500
Oil Shale	4.20 - 6.80	8.90 - 14.10[3]	30 - 300
Syncrude from Coal	4.70 - 7.60	10.30 - 16.20[4]	50 - 500
Fuels Requiring Automobile Modifications If Used as Stand-Alone Fuels			
Methanol from Coal	–	5.50 - 7.90	50 - 500[5]
Methanol from Biomass	–	8.20 - 14.60	50 - 500
Ethanol From Biomass	–	10.70 - 17.80	50 - 500

---

[1] Cost estimtes for synfuels may be low because commercial scale plants have not yet been built. The values given encompass currently accepted best estimates.

[2] In order to compare refined liquids (e.g., methanol and ethanol) with unrefined liquids (e.g., crude oil, shale oil, and syncrude), the following methodology is used. Where necessary (shale oil and syncrude), upgrading costs are added to the raw liquid costs. The cost per gallon of refined liquids is then assumed to be 1.64 times the cost per gallon of the upgraded raw liquid, which is the current ratio of the cost of refinery gate regular unleaded gasoline and the average crude oil acquisition cost.

[3] Raw liquid cost of $25 - $40/bbl. plus $3.50 - $5.00/bbl. for upgrading.

[4] Raw liquid cost of $28 - $45/bbl. plus $5.00 - $7.00/bbl. for upgrading.

[5] This is not additive to the potential of syncrude from coal.

SOURCE: OTA, K.A. Rogers and R.F. Hill, Coal Conversion Comparison, prepared for U.S. Department of Energy under contract EF-77-C-01-2468, and Coal Liquids and Shale Oil as Transportation Fuels, A Discussion Paper of the Automotive Transportation Center, Purdue University, West Lafayette, Indiana, July 6, 1979.

oil shale or coal liquification plants of comparable capacity. And comparable investments or subsidies designed to encourage increased conservation and enhanced oil recovery could yield much larger supplies of liquid fuel. Although an assessment of the alternatives has not been conducted, these are important questions which can influence the desirability of fuel ethanol production in the 1990's.

Although synfuels from coal and shale are expected to be produced during the 1990's, atmospheric build-up of CO_2 could alter these plans. If CO_2 becomes an overriding concern, ethanol from crop residues and wood would become much more attractive.

Until the uncertainties are resolved, however, investment in ethanol distilleries is likely to be limited to total production levels below that which is physically possible and economically viable in the 1980's.

III. ENVIRONMENTAL EFFECTS

Perceptions about the environmental benefits and costs of gasohol have focused on the potential air quality effects of emissions from gasohol-powered automobiles. Each stage of the gasohol "fuel cycle" has significant environmental effects, however, and the most important effects are likely to be the result of growing and harvesting the ethanol "feedstocks" - starch and sugar crops, crop residues, grasses and wood.

OBTAINING THE FEEDSTOCK

Starch and sugar crops would be the most likely near-term candidates for the ethanol feedstocks of a large-scale gasohol program; proven conversion technologies exist for these crops, and large acreages suitable for conversion to intensive agriculture are currently available. At the present time, pressure to promote gasohol is stressing the use of surplus and distressed crops as well as food wastes, but supplies of these feedstocks are limited. A commitment to produce quantities of gasohol greater than these sources can provide (i.e., more than a few hundred million gallons of ethanol per year) must involve additional crop production through more intensive cultivation of present cropland and the development of "potential" cropland currently in forest, range or pasture. A commitment to produce enough gasohol to supply most U.S. automotive requirements could involve putting approximately 30-70 million additional acres into intensive crop production. Assuming the acreage was actually available, this new crop production would accelerate erosion and sedimentation, increase pesticide and fertilizer use, replace unmanaged with managed ecosystems, and aggravate other environmental damages associated with American agriculture.

Soil erosion and its subsequent impact on land and water quality will be a significant impact of an expansion of intensive agricultural production. Current agricultural production is the primary cause of soil erosion in the U.S.: between 2 and 3 billion tons of soil from American farms enter the nation's surface waters each year. (28) The soil particles cause turbidity, fill reservoirs and lakes, obstruct irrigation canals, and damage or destroy aquatic habitats. In addition, they transport other water pollutants including nitrogen, phosphorus, pesticides, and bacteria. (28) Although the extent of the damage to aquatic ecosystems is unknown, yearly material damage from sedimentation has been estimated at over $1 billion.(29) Aside from damages associated with these water impacts, allowing a sustained soil loss of more than about 5 tons/acre year eventually will rob the land of its topsoil. Average erosion rates on intensively managed croplands currently exceed these levels by a considerable margin. For example, sheet and rill erosion alone on intensively managed croplands averages 6.3 tons/acre year nationally and 7.3 tons/acre year in the Corn Belt. (30) These high rates of erosion are allowed to persist because in all but the most severe cases the loss of valuable topsoil is slow. A net loss of 10 tons/acre year leads to a loss of only an inch of topsoil in 15 years. Depending on the depth of the topsoil and the depth and quality of the subsoil, the loss in productive potential over this length of time may be significant or negligible. Even a significant loss may go unnoticed, because it is masked in the short term by productivity improvements resulting from improvements in other farming practices or more intensive use of agricultural chemicals. Eventually, however, continuing losses in productive potential could cause a leveling off and even a decline in U.S. farmland productivity.

Erosion from current production appears to be a reasonable model on which to base evaluations of future erosion potential from ethanol crop production. An examination of Soil Conservation Service land capability data indicates that the lands most likely to be shifted to intensive ethanol feedstock production are somewhat more erosive than land that is currently being cultivated, but not excessively so. On a national basis, 48% of the land in intensive crops is erosive compared to 53-60% of the land that is most likely to be shifted to intensive production. (30) Although precise data are not available, the land currently set aside probably would be both the first to be used and the most erosive of the land base for biomass energy crops.

The extent of any erosion problem will depend on the type of crops grown. In general, annual crops are more erosive than perennials, and row crops more than close-grown crops. Thus, corn (an annual row crop), the most widely discussed gasohol crop, would be among the most erosive; forage grasses (perennial close-grown crops) may be among the least.

A large expansion in intensively managed cropland will have important impacts in addition to erosion damage. For example, pesticide use -- currently at about one billion pounds per year for the U.S. (29) -- will expand somewhat proportionately to the expansion in acreage. Although the long-term effects of pesticides are not well understood, some pesticides (e.g., Aldrin, Dieldrin, Mirex) have been banned from use because of their potential to cause cancer or other damage -- and it is possible that other widely-used pesticides will be discovered to be dangerous as more knowledge accumulates. Public interest in pesticide dangers to human health --

whether proven or merely perceived -- appears to be sharply on the rise. OTA considers it a strong possibility that public reaction to health damages reported to be linked to pesticide use may increase dramatically in the future. This may constrain both the continuing rise in pesticide usage and the expansion of crop production for energy feedstocks.

Another important issue concerns the heavy use of fertilizers on new cropland. Fertilizer application rates on this land probably will be high because the payoff in increased yield is well established. Runoff and leaching of nutrients to surface and groundwaters will cause premature aging of streams and other damage to aquatic ecosystems. In addition, natural gas must be used to produce nitrogen fertilizers for the new crops (or to replace the nitrogen embodied in the residues removed). At current application rates, 50 million acres of corn production would require over 100 billion cubic feet of gas per year, or over 1/2 of 1% of total U.S. natural gas production.

The increase in cropland also would involve a transformation of unmanaged or lightly managed ecosystems -- such as forests -- into intensively managed systems. For example, approximately one quarter of the land identified by USDA as having a high or medium potential to become cropland is forest, (31) and the Forest Service considers this land -- especially in the Southeast -- as a prime target for conversion. A full-scale national gasohol program could increase the pressure to clear as many as 10 to 30 million acres of unmanaged or lightly managed forest.

All of the impacts associated with increased crop production are functions not only of the type of crops grown but also of land capability, production practices, improvements made to the land, and other factors. There is enough freedom of choice in the system to significantly reduce the environmental impacts of a major gasohol program. Aside from choosing the land to be cultivated as well as the crop and tilling procedure, farmers may use a variety of environmental protection measures such as integrated pest management procedures, soil analysis to minimize fertilizer applications, and the development of disease-resistant crops to reduce impacts. The Environmental Protection Agency (through its 208 areawide planning process to control nonpoint sources of pollution) and the Department of Agriculture (through the Soil Conservation Service programs) have made only limited progress, however, in shifting farming practices toward more environmentally benign and soil conserving methods. (32, 33) Also, there is considerable controversy surrounding the net environmental effects and the potential impacts on crop yields of some of the measures advocated as environmentally beneficial. For example, conservation tillage, advocated as an extremely effective erosion control, requires increased applications of herbicides and insecticides (34) (the latter to combat insects that are sheltered by crop residues left on the surface as a mulch). Loss of these pesticides to surface waters will be slowed by lessening erosion, but increased contamination of groundwater may still result. Similar ambiguities, especially about the possibility of lowered net yields, surround measures such as pest "scouting" (monitoring), organic farming procedures, and other practices.

In light of farmer resistance to controls, the apparent lack of high priority given to most agricultural environmental problems by the EPA, and the possibility that certain environmental measures may replace one adverse environmental impact with another (for example, conservation tillage replacing erosion with increased herbicide use), OTA concludes that the environmental effects of converting tens of millions of acres to intensive production may be at least as great as the effects observed on similar acreage today.

Although food crops currently may represent the most economic ethanol feedstock, the potential for substantial increases in corn (and other sugar/starch crop) prices and for improvements in conversion processes for alternative feedstocks points to the eventual primacy of these alternative feedstocks in ethanol production. The use of crop residues, forage grasses, and other alternative feedstocks will have environmental consequences that are substantially different from those caused by growing and harvesting sugar/starch crops.

Crop residues may be used either as an ethanol feedstock or as a distillery boiler fuel. Although leaving crop residues on the surface is an important tool for erosion control, substantial quantities can be removed from flatter, less erosive soils in some parts of the Corn Belt and elsewhere without causing erosion to exceed 5 tons/acre year. (35) Also, many farmers plow these residues under in the fall to prevent them from harboring crop pests or to allow an earlier spring planting, thus losing their protection anyway. Thus, the use of residues will cause additional erosion only if they otherwise would have been left on the surface, and only if they are removed from erosion-prone lands or in excessive quantities. Unfortunately, conflicts between short-term profits and long-term land protection could easily lead to improper use of residues unless effective institutional controls or incentives for environmental protection can be developed. Also, there is some concern (although little substantive evidence) about possible harmful effects of reductions in soil organic levels caused by residue removal.

The intensive cultivation of forage grasses would cause pollution effects from fertilizers and pesticides, but could be expected to produce far lower levels of erosion than food crops (as noted above).

The major factor controlling the impact of these alternative feedstocks will probably be the efficiency with which they can be converted to ethanol. A breakthrough in conversion efficiency could nearly double alcohol production per ton of feedstock and halve the acreage -- and impacts -- necessary to sustain the desired gasohol use.

ETHANOL PRODUCTION

Significant environmental effects of ethanol production are associated with its substantial energy requirements and the disposal of distillation wastes.

New energy efficient ethanol plants probably will require about 50,000-70,000 BTU per gallon of ethanol produced to power the distilling, drying and other operations. Individual distilleries of 50 million gallons/year capacity will use as much fuel as 50-70 MW powerplants; a 10 billion gallon per year ethanol industry will use about the same amount of fuel as needed to supply 10,000-14,000 MW of electric power capacity.

New Source Performance Standards have not been formulated for industrial combustion facilities, and the degree of control and subsequent emissions are not predictable. The most likely fuels for these plants will be coal or biomass (crop residues), however, and thus the most likely source of problems will be their particulate emissions. Coal and biomass combustion sources of the size required for distilleries -- especially distilleries designed to serve small local markets -- must be carefully designed and operated to avoid high emission levels of unburned particulate hydrocarbons (including polycyclic organic matter). (36) Fortunately, most distilleries will be located in rural areas, and this will reduce total population exposure to any harmful pollutants.

The effluent from the initial distillation step -- called "stillage" -- is very high in biological and chemical oxygen demand and must be kept from entering surface waters without treatment. The stillage from corn and other grains is a valuable feed byproduct and it will be recovered, thus avoiding this potential pollution problem. The stillage from some other ethanol crops is less valuable, however, and may have to be strictly regulated to avoid damage to waterways. Control techniques are available for the required treatment.

If fermentation and distillation technologies are available in a wide range of sizes, small scale on-farm alcohol production may play a role in a national gasohol program. The scale of such operations may simplify water effluent control by allowing land disposal of wastes. On the other hand, environmental control may in some cases be more expensive because of the loss of scale advantages. Also, the current technology for the final distillation step, to produce anhydrous alcohol, uses reagents such as cyclohexane and/or ether that could pose severe occupational danger at inadequately operated or maintained distilleries. Although alternative (and safer) dehydrating technologies may be developed, in the meantime special care will have to be taken to ensure proper design, operation and maintenance of these smaller plants.

The decentralization of energy processing and conversion facilities as a rule has been viewed favorably by consumer and environmental interests. Unfortunately, a proliferation of many small ethanol plants may not provide a favorable setting for careful monitoring of environmental conditions and enforcement of environmental protection requirements. Regulatory authorities may expect to have problems with these facilities similar to those they run into with other small pollution sources. For example, the attempts of the owners of late model automobiles to circumvent pollution control systems conceivably may provide an analog to the kinds of problems that might be expected from small distilleries if their controls prove expensive and/or inconvenient to operate. Congress should carefully weigh the potential costs and benefits of centralized vs. decentralized ("on-farm") plants before providing incentives that might favor one over the other.

GASOHOL USE

The effects of gasohol use on automotive emissions are dependent on whether the engine is tuned to run fuel rich or lean and whether or not it has a carburetor with a feedback control. Although some gasohol advocates have claimed that the emissions effects are strongly positive, in fact it is difficult to assign either a beneficial or detrimental net pollution effect to gasohol use.

Gasohol use will have the following effects on most cars in today's automobile fleet (i.e., no carbureter modifications are made and fuel "leaning" takes place): (9)

o   increased evaporative emissions (although the new emissions are not particularly reactive and should not contribute significantly to photochemical smog)

o   decreased emissions of carbon monoxide

o   increased emissions of aldehydes (which are reactive and conceivably may aggravate smog problems)

o   increased NOx emissions with decreased emissions of exhaust hydrocarbons, or decreased NOx with increased HC (depending on the state of engine tune).

The emissions effects on automobiles which are manually or automatically adjusted to maintain constant air/fuel ratios (i.e., no "leaning" effect) will be considerably less.

This mixture of observed emissions reductions and increases, and the lack of extensive and controlled emissions testing, does not justify making a strong value judgement about the environmental effect of gasohol used in the general automobile population (although the majority of analysts have concluded that the net effect is unlikely to be significant). It may be possible to engineer an unambiguously beneficial effect, however, by channeling gasohol to certain urban areas with specific pollution problems (for instance, high carbon monoxide concentrations but no smog problems) or to vehicle fleets with engine characteristics that could maximize potential benefits from gasohol. The federal government could stimulate this type of use by initiating federal fleet use as an example, and by providing economic or regulatory incentives to fleet operators or to areas that would benefit from gasohol use.

GLOBAL EFFECTS OF THE GASOHOL FUEL CYCLE

The emission of carbon dioxide (CO_2) has become a major issue in the debate over synthetic fuels production.

Net CO_2 emissions from the gasohol fuel cycle are dependent on the extent and nature of land conversion needed to grow the feedstock, the fuel used to fire the distilleries, overall energy efficiency of the fuel cycle, and the type of fuel displaced (gasoline from natural crude or gasoline from coal-derived synfuel). If a minimum of forested land is permanently cleared for growing ethanol crops, if the major distillery boiler fuel is crop residues or some other renewable fuel, and if the ethanol is efficiently used (as an octane booster), then universal use of gasohol will reduce current CO_2 emissions from automobile travel by about 10%*.

It should be stressed, however, that even maximum use of alcohol fuels in the U.S. can have only a small effect on total worldwide CO_2 emissions. A combination of major changes in the current energy system and a significant slowdown of deforestation, effected on a worldwide scale, would probably be needed to put a brake on increasing atmospheric CO_2 levels.

---

* One uncertainty in this conclusion is the extent to which organic loss on cultivated land is an important CO_2 source.

IV. SOCIAL IMPACTS

The widespread production and use of gasohol can be expected to have a number of social and economic effects. These include impacts that are more likely to be perceived as important at the local level (such as changes in employment, demography, public services, and quality of life) as well as impacts that can be national or international in scope (for example, changes in the economy, land ownership, institutions and politics and ethical considerations). Some of these impacts are quantifiable, while others can only be discussed qualitatively. It should be noted that the scope and magnitude of these effects are highly uncertain because no reliable methodology for predicting the social impacts of emerging technologies exists and because the size and location of projects are unknown. Consequently, this discussion will only be able to identify some of the potential social changes that could occur if gasohol were used widely.

LOCAL IMPACTS

Increased production and consumption of gasohol would create a variety of new jobs. Approximately 15-19 million additional hours of farm labor would be required to produce 1.3 billion gallons of ethanol per year from corn (0.1 quad/year). (37) (Comparable productivity estimates were not available for feedstocks other than corn.) Additional employment opportunities would arise in the transpoartation of feedstocks to distilleries and of ethanol to refineries or gasohol distributors, as well as in the manufacture and delivery of fertilizer, farm machinery, distillery equipment, and in the construction and operation of distilleries. Estimates of the number of distillery operating, maintenance, and supervisory personnel required to produce 1.3 billion gallons of ethanol per year from corn range from 1,200 to 4,000. (5, 20) Comparable figures were not available for distillery construction or for the manufacture of distillery equipment.

The production of distillery fuels would also create employment opportunities on farms or in coal mines. The use of crop residues and/or cellulose crops to fire distillery boilers would require additional farm labor, but not on the same scale as would the production of corn for ethanol feedstocks. For example, harvesting corn residues requires 6-10 million work-hours per 1.3 billion gallons of ethanol for a large round bale system, or 3.5-4.5 million work-hours per 1.3 billion gallons of ethanol for a large stack system. (10) Labor requirements for harvesting collectible residues and moving them to the roadside are summarized in Table 5. Additional labor would be required to transport the residues to a distillery. Alternatively, if distilleries are fueled with coal, approximately 375,000-600,000 underground coal mine worker shifts or 125,000-200,000 surface mine worker shifts would be required to produce 1.3 billion gallons of ethanol. (38)

TABLE 5

LABOR REQUIREMENTS FOR HARVESTING COLLECTIBLE RESIDUES

(million work-hours/1.3 billion gallons)

	Large Round Bales	Large Stacks
Corn	6.2-9.7	3.5-4.4
Small Grains	3.8-6.8	3.4-4.1
Sorghum	14.4-15.2	10.0-10.3
Rice	14.8	
Sugar Cane	11.2	

Source: Reference 10.

57

For the most part, the employment opportunities discussed above represent the creation of new jobs rather than the transfer of existing agricultural and energy sector employment to the production of gasohol. To the extent that current food and feed production is used for ethanol, however, new jobs are not created. In addition, use of corn stillage as animal feed would compete directly with soybean meal and may reduce employment in that industry. (10)

It should be noted that estimated labor requirements in agriculture are very uncertain. Crop production is highly mechanized and labor requirements have declined continuously since 1950. If farm labor productivity continues to increase, the estimates given above are high. Other uncertainties are introduced by the projected method of increasing production; more labor usually is required to expand the number of acres in production than to increase the output per acre, and some crops require more labor than others.(10) In addition, during peak farm seasons such as planting and harvesting, agricultural labor often is scarce. Emphasizing crops that require less intensive management and that are harvested at different times of the year from conventional food and feed crops could alleviate this problem.

The impacts of new employment opportunities depend in part on where they arise and in part on whether they are filled by residents or in-migrants. The eastern half of the U.S. has the greatest amount of potential cropland. (31 Assuming that these lands are used to produce ethanol feedstocks, most of the employment opportunities associated with gasohol would arise in these regions. Productivity on some of the lands in the North, however, ultimately is limited by water availability, climate, and other factors. Thus, increased production in the North probably will be greatest in the Corn Belt and Lake States.

On-farm employment and new jobs asociated with distillery operation (except for jobs requiring special skills) probably would involve long-term rural residents. This could reduce off-farm migration, shift the age distribution in rural areas to a younger population, revitalize small farming communities, and increase the demand for migratory workers during harvest season. On the other hand, distillery construction is more likely to involve temporary in-migrants or commuters. Rural agricultural areas are not well equipped to accommodate in-migrants, and temporary shortages of housing, education and medical facilities, and other public and private sector goods and services could occur during construction. These impacts will be minor, however, in comparison to those associated with energy development in the West. Although a distillery would contribute significant amounts to the local tax base, tax revenues usually do not begin to accrue until a facility is in operation.

NATIONAL AND INTERNATIONAL IMPLICATIONS

In addition to increases in tax revenues, ethanol production could have significant economic impacts on the price of food and farmland. Should the demand for ethanol feedstocks increase more rapidly than the supply, the result would be increases in farm commodity prices and farm income. Many agricultural economists believe that this situation leads to increases in farmland prices that permanently increase the cost of farming. Although this would benefit the landowner, it also would threaten the viability of farming on rented land by eliminating gains in farm income. In addition, it could endanger small farmers' ease of access into the market and accelerate the trend toward large corporate holdings of farmland.

Increases in the cost of farming and competition for ethanol feedstocks between energy and food markets could also increase the cost of food. This increased cost falls disproportionately on the poor. In addition, increases in U.S. food prices are likely to increase the cost of food on the international market. Some countries will not be able to afford food imports, and others will export crops now used domestically for food. In either case the net result would be to worsen the world food situation. It is not known, however, at what level of increase in land and food prices these effects will occur, and their final impact cannot be determined.

The institutional impacts of increased gasohol use include changes in governmental and social structures and in attitudes and public opinion as well as ethical considerations. Within the government, the principal changes would occur in federal and state agencies. The use of farm commodities and currently unproductive cropland to produce ethanol would require the Departments of Agriculture and Energy to cooperate on both energy and agricultural policy. Changes in existing tax policy also may be necessary to facilitate the production of ethanol for fuel, and to prevent the loss of tax revenues that normally would accrue from sales of gasoline and alcoholic beverages.

Changes in social institutions probably would evolve over longer periods of time. Increases in employment on farms and in rural agriculural areas would decrease the number of young people leaving these areas and ultimately strengthen the rural family and farming as a way of life. On the other hand, significant population increases in rural areas with "one-crop" economies could result in impacts that would destroy long term residents' sense of community and rural quality of life.

Favorable individual and group attitudes and general public opinion are politically and practically necessary to large-scale production of ethanol. Favorable public opinion is politically necessary for the funding and implementation of government programs directed toward the supply of and demand for gasohol. Favorable attitudes among farmers toward the conversion of currently unproductive land to ethanol feedstock production are also necessary if these programs are to be effective. Although the use of agricultural lands for ethanol feedstocks is likely to be politically popular among most non-agricultural groups, the conversion of non-productive federal land (for example, Bureau of Land Management lands) to cropland probably would be opposed by some interest groups, such as conservationists.

In addition, favorable attitudes in the farm sector toward the production and use of ethanol fuels will be necessary to the success of small on-farm systems. Recent research on the adoption of innovations in agriculture suggests that the best predictors of the adoption of commercial (as opposed to environmental) innovations are above-average farm capital, size and sales, as well as the farmer's education. These findings were correlated with traditional agricultural extension service strategies for the voluntary adoption of innovations by farmers. (39) These strategies are based on the well-documented diffusion process of commercial practices, and probably can be applied to the on-farm production and use of ethanol.

Finally, the increased use of gasohol can raise ethical considerations related to the conflict between food and feed, on the one hand, and energy on the other. This conflict has become increasingly prominent in the last decade as both food shortages and the finiteness of conventional energy resources have become recognized as world problems. In the U.S., this conflict historically has revolved around the use of prime agricultural land for surface and, to a lesser extent, underground coal mines, and around energy uses of water in the arid regions of the West.

Increased demand for farm commodities to be used for domestic fuel will highlight this conflict because fuel use will compete directly with U.S. food and feed exports. If food exports are reduced significantly in order to augment U.S. energy supplies, adverse foreign responses could result. The use of farm commodities for ethanol also will compete directly with domestic consumption of food and feed, and limits on the sale of commodities for energy could become necessary. In addition, if ethanol feedstock suppliers are given long term guarantees in order to stimulate production of gasohol, and if the demand for food continues to rise, Americans ultimately could be forced to choose between relatively inexpensive food and relatively inexpensive fuel.

Of the social and economic impacts discussed above, those that are most likely to become problems include potential increases in farm commodity and farmland prices, and potential conflicts between the use of commodities for energy rather than food or feed. The timing and magnitude of increases in the price of food and land cannot be determined, however, and their total effect is uncertain. The production and pricing of ethanol feedstocks could be integrated into overall U.S. agricutural and energy policy before these impacts become severe. Other long-term social and economic impacts of gasohol production and use -- revitalization of farm families and rural communities, as well as increased domestic energy self-sufficiency -- would be beneficial.

V. CURRENT FEDERAL PROGRAMS AND POLICIES

INTRODUCTION

The responsibility for gasohol development is spread among a number of federal agencies having different duties as well as contrasting, and in some instances conflicting, perspectives on gasohol. No clearly focused or operational federal policy on gasohol development has appeared to exist. The thrust of the federal government's efforts has typically been to respond to Congressional initiatives.

In FY 1979, OTA estimates that federal expenditures of between $13 and $17 million directly supported the development of alcohol fuels from biomass. In FY 1980 the Administration's research activities are expected to be funded at a level between $18 and $25 million. Additional subsidies include $40 million in loan guarantees, exemption of the federal excise tax on gasohol (for domestic production and imports), eligibility of alcohol fuels for entitlement awards, and an investment tax credit of 20% on alcohol fuels facilities. Well over 90% of the federal government's cumulative expenditures (since 1975) have accrued in the last year.

The Department of Energy is the lead agency responsible for formulating energy policy and for the development of alcohol fuels technology. DOE's responsibilities overlap the Department of Agriculture's responsibility to administer food and fiber production programs, and although a number of other agencies are involved (including the Departments of Commerce and Treasury), DOE and USDA are the principal agencies with jurisdiction over alcohol fuels development.

USDA POLICY

The USDA has been involved in agriculture policy since the Federal Farm Board was established in 1929. The Department has typically relied on carrot-and-stick combinations of supply control and price support programs to insure market stability, income protection to producers, and food security. These programs, together with extensive research and development programs (USDA FY 1980 solar energy R&D expenditures total more than $27 million), place USDA in a unique position to develop biomass.

USDA policy towards gasohol development has historically lacked clear or consistent direction. Many agencies and programs within the Department have sometimes advocated conflicting or contradictory positions on gasohol. The Department has characteristically been in the situation of reacting to gasohol initiatives proposed by the Congress rather than developing or implementing their own.

The current thrust of USDA gasohol policy is to take a wait-and-see approach towards new or more energetic gasohol initiatives, other than those already proposed and enacted by the Congress. The Agency has emphasized R&D rather than implementation on the premise of resolving technical and policy uncertainties before implementing broad-scale programs having many unknown impacts. It is the view of USDA that Congress has already provided the agency with sufficient authority, particularly in the 1977 and 1978 Agricultural Acts, to support alcohol fuels production. USDA does not advocate further expansion of programs or policies until it is clear that additional initiatives are warranted.(23)

USDA's current stance toward alcohol fuels is that agricultural policies and programs, and an alcohol fuels industry, can be mutually supportive only in various incidental or accidental ways. USDA further explains that (a) agriculture price support and stabilization policies serve different functions than any alcohol fuels program, and therefore no substitution and shifting of outlays is sensible; (b) due to extreme uncertainty in cropland availability, any commitment to a grain-based ethanol program should be restricted, in order to retain the options of foregoing further commitment, or of withdrawing completely, to minimize unrecoverable costs.

USDA R&D

No agency in the federal government has more abundant resources or greater administrative capability to research and implement biomass and alcohol fuels than does the USDA. Alcohol fuels development is so intertwined with food policies and the agricultural sector that in many cases USDA's role in its development is essential. Biomass energy R&D, however, has a low overall priority in USDA's research, in part because the Agency has no real energy mandate. The agency has a tendency to avoid burdening its (declining) research budget by funding work relating solely to energy.

Although alcohol fuels R&D has been emphasized over implementation-oriented activities, alcohol fuels have received relatively little attention. In FY 1978, of the $6 million biomass budget, little supported alcohol fuels development directly. In FY 1979, about $1 million supported alcohol R&D (out of a total biomass budget of $9 million). In FY 1980, it is expected that somewhere between $4 million and $6 million will be divided approximately equally between alcohol feedstock production and advanced conversion systems (of a total biomass budget of almost $24 million). (The ranges in the budgets reflect the department's uncertainty in determining expenditures.)

Biomass and alcohol fuels R&D in the Department of Agriculture have, with some exceptions, suffered from a lack of direction and poor coordination(40). For example, USDA has made it a goal to aid the agriculture sector in becoming net-energy self-sufficient by 1990(41). The Agency has not, however, developed any plans, nor is it following any specific research strategies, to develop alcohol fuels or any other energy applications in the agriculture sector. Although this is due in part to the low emphasis given to alcohol fuels, it is also a reflection of USDA's highly decentralized management with historically well defined areas of responsibility. A great deal of management responsibility for research resides in field offices and land-grant institutions, and it can be difficult to direct R&D policy under these circumstances. Newly established biomass/alcohol fuels programs in FY 1980 may alleviate these sorts of difficulties, but they will not change fundamental management problems.

DOE POLICY

The Department of Energy is the lead agency responsible for developing alcohol fuels technologies. In the past, DOE has focused its efforts in the area of R&D; little emphasis was given to commercialization. Recently, however, DOE has taken a somewhat more active role in supporting the technology's near-term development. Its role has been more aggresive than that of USDA.

DOE's current policy is designed to achieve low to moderate levels of ethanol production in the near term. The Agency is relying predominantly on two federal incentives: (1) exemption of the federal excise tax on gasohol blends, worth $16.80/barrel, and (2) entitlement awards to alcohol fuels worth 2 to 3 cents/gallon, or roughly $1.00/barrel. With these subsidies DOE projects that ethanol production can reach 500 to 600 million gallons annually by 1985 using wastes (e.g., cheese whey) rather than agricultural commodities. The Agency at this time does not support any significant expansion of programs or subsidies to further stimulate the production of gasohol. (It should be noted that the Administration has recently proposed several synthetic fuels initiatives that are projected to achieve ethanol production levels of over 1 billion gallons annually. It is unclear at this time, however, whether these levels can actually be reached with the initiatives proposed.)

DOE R&D

DOE has responsibility for molding the federal government's alcohol fuels research effort. In FY 1978, DOE expenditures for alcohol from biomass R&D totaled almost $5.5 million. In FY 1979 and FY 1980, OTA estimates that alcohol related expenditures will total $12 to $14 million and $14 to $17 million respectively. During these years, 50% to 65% of program funds supported conversion R&D, 25% to 35% supported end-use studies, and 15% to 25% went towards production and collection research.

Overall, DOE's technology development efforts on conversion technologies have been balanced and supportive of a range of promising low- to high-risk technical options. The same is true to some extent in the area of biomass production and collection. OTA has found that DOE biomass programs have supported a narrow range of long-term technological applications, but this has not been the case in the specific area of alcohol fuels development. Whereas OTA has determined that many DOE biomass programs have been fragmented and administered ineffectively, management problems do not appear to have substantially affected the Agency's alcohol fuels development efforts(40).

INTERELATIONSHIP BETWEEN DOE AND USDA

The Department of Energy is designated as the lead agency in bioenergy R&D and as such has responsibility to integrate and coordinate alcohol fuels technology development. At the same time, USDA is responsible for administering agricultural production policies, as well as R&D. Since alcohol production and use is intertwined with the farm and energy sectors, the success of an alcohol fuels development program is in part contingent upon the implementation of complimentary production, conversion, and end-use policies and research programs by the two agencies.

The agencies, however, have made few efforts to integrate agriculture and energy policies. A comprehensive framework has not been established to perform this role adequately, and inter-agency coordination has been poor. In the area of research, the coordination of DOE and USDA biomass programs has been improving. Inter-agency coordination of alcohol fuels programs has not, however. If coordination is to improve, administrative and technical differences between these two very different agencies must be resolved(40).

RESEARCH AND DEVELOPMENT NEEDS

OTA has identified major areas where research and development seem particularly important. The purpose of this section is to describe these research needs and to indicate to what degree the federal research effort is currently addressing them. It should be understood that not all possible research areas should necessarily be addressed by the federal government.

Feedstocks

Develop feedstock crops with higher yields of ethanol per acre. Comparative studies of starch and sugar crops, and high-yield hybrids, which are candidate feedstocks. Regional studies should examine productivity as a function of soil type, weather, etc. (The federal government directly and indirectly, has supported a great deal of research in this area. The research, however, has focused on food and feed production rather than on energy production, and research is needed to assimilate the existing data.)

Investigation of the effect of uncertainty and variability of prices and supplies in the agriculture markets on the potential of producing energy from agriculture. (The Federal government has supported a very limited and narrow range of investigations in this area.)

Develop ethanol feedstock crops which are nitrogen fixing, so as to reduce the energy inputs to farming. (USDA is supporting a significant amount of research in this area.)

Investigate the feed value of distillers' grain, particularly at high levels in the feed and with large water content. (USDA is supporting some research in this area.)

Develop nitrogen fixing bacteria which can be substituted for nitrogen fertilizer. (This is a basic research area which NSF has supported to some degree.)

Screening of unconventional plant types as candidate feedstocks. (The federal government has supported little research in this area.)

Regional studies to evaluate the availability of residues. Detailed evaluations need to be performed to determine how residue use can alleviate and/or exacerbate environmental problems such as soil erosion. Analyses of institutional constraints of residue use are also needed. (USDA has supported research in this area. Institutional issues, however, have not been addressed.)

Environmental effects of increased forage grass production. (The federal government has sponsored little research in this area.)

Investigation of the environmental effects of pesticides, herbicides, fertilizers, soil erosion, and other effects associated with increased agricultural production. (USDA has supported a limited amount of research in this area, but litle has been done regarding energy production.)

Analysis of the availability and productivity of potential crop lands, the costs of bringing this land into production, and its effect on agriculture markets. (Little research has been done in this area.)

Impacts of protein (e.g., distillers' dried grains) on conventional (and non-conventional) domestic and international markets. (Little research has been supported in this area.)

(There are many R&D problems associated with obtaining feedstocks from the forest sector which are not mentioned here but which could greatly influence feedstock availability.)

Conversion

Research into improving the yields of cellulose hydrolysis. (DOE and NSF are supporting research in this area.)

Basic thermochemical research into rapid pyrolysis, particularly to attain high ethylene yields. (The federal government has supported little research in this area.)

Application of solar-thermal systems to distillation. (The federal government is supporting little research in this area.)

Development of simple and less energy intensive methods for concentrating ethanol-water mixtures, e.g., phase separating salts, vacuum distillation, absorption processes, desiccants, freeze crystalization, membrane applications, and extraction. (The federal government is supporting research in these areas to a very limited degree.)

Develop low-cost methods to produce anhydrous ethanol in small-scale applications. (The federal government has supported little or no research in this area.)

Definition of the environmental effects of distilleries. (The federal government has sponsored a limited amount of research in this area. Little analysis, however, has been done on on-farm systems.)

Develop on-farm distilleries which are relatively automatic. (USDA and DOE are beginning some research in this area.)

Develop continuous fermentation processes. (DOE is sponsoring research in this area.)

Gasification of biomass as a source of hydrogen. (Present source of hydrogen -- used to produce fertilizer -- is natural gas.) (The federal government is sponsoring little research in this area.)

Examine energy and non-energy uses of liquid by-products. (A limited amount of research is funded by NSF.)

Resarch on the conversion of hemicellulose to liquids.
(DOE is sponsoring a limited amount of research in this area.)

End-Use

Determine the thermal efficiency of gasohol, in terms of best fuel blend. (DOE has recently begun work in this area.)

Develop emulsions and additives which can eliminate the need for using dry ethanol. (The federal government has supported litle research in this area.)

R&D on the use of pure alcohols (and applicable lubricants) in the motor fleet. (DOE has work on-going in this area.)

Determine long-run effects on performance, efficiency, and materials compatibility associated with the use of gasohol. (DOE has funded a limited amount of work in this area.)

Study of the effect of net increased aldehyde concentrations on air pollution. (DOE has begun some research in this area.)

Studies on the combustion chemistry of gasohol. (The federal government has supported research in this area.)

Examination of the potential for using gasohol in specific regions where its use may have unambiguously positive results. (The federal government has sponsored no research in this area.)

Research various gasoline compositions to determine ways of effectively using ethanol's octane boosting properties while minimizing evaporative emissions. (DOE has sponsored a limited amount of research in this area.)

Field test phase-stability of gasohol in distribution systems. (Little research has been done in this area.)

Evaluation of the long-run effects of using alcohol in diesels. (Little research has been done in this area.)

REFERENCES

1. OTA contractor report, <u>Agriculture and Agricultural Product Processing</u> <u>Wastes.</u>

2. <u>The Report of the Alcohol Fuels Policy Review</u>, Asst. Secretary for Policy Evaluation, U.S. Department of Energy, June 1979, Stock No. 061-000-00313-4, U.S. Government Printing Office, Washington, D.C.

3. Estimates based on conversations with Martin Andrias, President, Archer Daniels Midland, Decatur, IL; Prof. Wm. Scheller, Univ. of Nebraska, Lincoln, NE; U.S. Treasury and U.S. Department of Energy data; and others.

4. George Lovelock, President, American Gasohol, 223 Gerico Turnpike, Mineola, NY, private communication, August, 1979.

5. OTA contractor report, <u>Biological Production of Liquid Fuels and</u> <u>Chemical Feedstocks.</u>

6. Raphael Katzen Associates, "Multiple Product Waste Hardwood Facility - Ethanol, Furfural, Phenol", in The <u>Feasibility of Utilizing Forest Residues</u> <u>for Energy and Chemicals</u>, Report No. PB-258-630, Forest Products Laboratory, Madison, WI, March 1976.

7. Ralph Kienker, Monsanto Chemical Co., St. Louis, MO, private communication, July 1979.

8. OTA contractor reports, <u>Engineering Aspects of Thermochemical</u> <u>Conversion</u> and <u>Thermochemical Conversion of Biomass: The Scientific Aspects.</u>

9. OTA contractor report, <u>End Use of Fluids from Biomass as Energy</u> <u>Resources in Both Transportation and Non-Transportation Sectors.</u>

10. OTA contractor report, <u>The Potential of Producing Energy from</u> <u>Agriculture.</u>

11. Prof. Richard K. Pefley, Dept. of Mech. Eng., Santa Clara University, Santa Clara, CA, private communication, 1979.

12. See for example M.R. Ladisch and K. Dyck, Science <u>205</u>, p. 898 (1979).

13. For example, Robert Chambers, President, ACR Process Corp., Urbana, IL, private communication, September 1979.

14. William Davis, Bureau of Alcohol, Tobacco, and Firearms, U.S. Treasury, Washington, DC, private communication, July 1979.

15. Mr. Jacques Maroni, Energy Planning Manager, Ford Motor Company, Dearborn, MI, private communication, August 1979.

16. Dr. Robert Hirsch, EXXON Research and Engineering Company, Florham Park, NJ, private communication, July 1979.

17. Federal Energy Administration, U.S. Department of Energy, contractor report prepared by Gordian Assoc., Inc., <u>Energy Conservation, The Data Base,</u>

The Potential for Energy Conservation in Nine Selected Industries, Vol. 2, Petroleum Refineries, 1975.

18. William A. Scheller, Nebraska 2 Million Mile Gasohol Road Test Program, Sixth Progress Report, 1 April 1976 to 31 December 1976, Dept. of Chem. Eng., University of Nebraska, Lincoln, NE, January 31, 1977.

19. May 17, 1979 update of Domestic Crude Oil Entitlements, Application for Petroleum Substitutes, ERA-03, submitted to DOE by Archer Daniels Midland Co., Decatur, IL.

20. Raphael Katzen Assoc., Grain Motor Fuel Alcohol, Technical and Economic Assessment Study, December 1978, prepared for Asst. Secretary for Policy Evaluation, U.S. Department of Energy, Washington, DC, published June, 1979; Stock No. 061-000-00308-8, U.S. Government Printing Office, Washington, DC.

21. Profs. Carroll Bottum and Wallace Tyner, Dept. of Agricultural Economics, Purdue University, West Lafayette, IN, private communication, 1979.

22. Statement by Bob Berglund, Secretary of Agriculture, before the Committee on Science and Technology, Subcommittee on Energy Development and Applications, U.S. House of Representatives, May 4, 1979.

23. L.J. Barbe, Jr., Manager of Oil Movements, EXXON Pipeline Co., Houston, TX, private communication, August 1979.

24. Robert Reinstein, Energy Regulatory Administration, U.S. Department of Energy, Washington, DC, private communication, August 1979.

25. Preliminary estimates of the current farm program provided by USDA analysts.

26. OTA contractor report, Land Availability for Biomass Production.

27. OTA, Materials and Energy from Municipal Waste, Vol. 1, Stock No. 052-003-00692-8, U.S. Government Printing Office, Washington, D.C.

28. U.S. Environmental Protection Agency, Environmental Implications of Trends in Agriculture and Silviculture. Volume I: Trend Identification and Evaluation. EPA-600/3-77-121, October, 1977.

29. Soil Conservation Service, Draft Impact Analysis Statement: Rural Clean Water Program, June, 1978.

30. Based on computer runs conducted for OTA by the Soil Conservation Service.

31. Soil Conservation Service, 1977 National Erosion Inventory - Preliminary Estimates, Tables of Potential Cropland, April 1979.

32. General Accounting Office, To Protect Tomorrow's Food Supply, Soil Conservation Needs Priority Attention, CED-77-30, February, 1977.

33. President's Council on Environmental Quality, Environmental Quality:

Ninth Annual Report, December, 1978.

34. U.S. Environmental Protection Agency, Environmental Implications of Trends in Agriculture and Silviculture, Volume II Environmental Effects of Trends, EPA-600/3-78-102, December, 1978.

35. W.E. Larson, et al, "Plant Residues -- How Can They Be Used Best", Paper No. 10585, Sci. Journal Series, SEA-AR/USDA, 1979.

36. N. Dean Smith, "Organic Emissions from Conventional Stationary Combusion Sources", U.S. Environmental Protection Agency, Research Triangle Park, N.C., August 1977.

37. USDA, Agricultural Statistics, 1978, Washington, DC: US Government Printing Office.

38. OTA, The Direct Use of Coal. Washington, DC: US Government Printing Office, 1979.

39. Pampel, Fred, Jr., and J.C. van Es, Environmental Quality and Issues of Adoption Research. Rural Sociology, 42, No. 1, p.57 (1977).

40. OTA contractor report, Federal Bioenergy Programs, prepared by Mark Gibson, March 1979.

41. Statement of Weldon Barton, Director, Office of Energy, USDA, at the Mid-American Biomass Energy Workshop, Purdue University, May, 1979.

Printed in the United States
68199LVS00007B/243